DON'T BLAME
THE
WEATHERMAN

DON'T BLAME THE WEATHERMAN

Mr. G. Talks to You About the Weather

IRV "MR. G." GIKOFSKY

Illustrated by Wendy Miller

AN AVON C CAMELOT BOOK

DON'T BLAME THE WEATHERMAN is an original publication of Avon Books. This work has never before appeared in book form.

AVON BOOKS
A division of
The Hearst Corporation
1350 Avenue of the Americas
New York, New York 10019

Copyright © 1992 by Irv Gikofsky
Back cover author photograph by Brian Edwards
Interior illustrations by Wendy Miller
Published by arrangement with the author
Library of Congress Catalog Card Number: 92-20854
ISBN: 0-380-76720-1
RL: 5.6

Library of Congress Cataloging in Publication Data:
Gikofsky, Irv.
    Don't blame the weatherman : Mr. G. talks to you about the weather / Irv Gikofsky.
        p.   cm.
    1. Weather forecasting.   I. Title.
QC995.G47   1992   92-20854
551.6′3—dc20   CIP

First Avon Camelot Printing: November 1992

CAMELOT TRADEMARK REG. U.S. PAT. OFF. AND IN OTHER COUNTRIES, MARCA REGISTRADA, HECHO EN U.S.A.

Printed in the U.S.A.

OPM   10   9   8   7   6   5   4   3   2

Acknowledgments

Twenty-three years ago I met Elliot Shapiro. I knew of him as perhaps Harlem's best and most compassionate principal from reading the book *Our Children Are Dying*, detailing his work with the children of P. S. 119. I first saw him in the documentary, "Marked for Future," and was deeply touched by his sensitivity and insight. However, it was only after I first met him, when I invited Elliot and his wife Florence to my college class as a guest speaker, that I realized the book and film fell far short of the man. Dr. Shapiro is a combination of gentle and tough, idealist and realist, humble and shy, yet fights fiercely for the people in distressed areas and especially the children of the poor. For the last twenty-three years he has been my mentor and hopefully I've been in some measure his protégé. To his indomitable spirit, to his reservoir of goodwill, to his nurturing kindness, and above all, to his unconditional support and love, I dedicate this book. Dr. Shapiro has been described by Dr. Stan Goldsmith of the Sloan-Kettering Institute as "a national treasure." Elliot is both that and my dearest and best friend.

Contents

1

Sunny Until Further Notice: Forecasting the Weather

Why do people always talk about the weather?
People always talk about the weather because it's always there! From the moment you get up in the morning until the moment you go to sleep (and even while you're sleeping), you're affected by the weather. Will it be hot, will it snow, will it rain on my parade? These are the questions people have been asking about the weather since man has been on earth. And now, with the technology available to us we can—and need to—know what the weather will be. Another reason that people love to talk about the weather is because it's a "safe" topic about which no one should disagree!

Who predicts the weather?

There are plenty of amateur weather forecasters, but the weather you hear about on the radio or TV is probably predicted by a meteorologist. Meteorology, by the way, is the study of weather as a science. This includes weather that was in the past as well as what's to come in the future. By tracking past weather patterns, we can sometimes predict what may be in our weather future.

The National Weather Service has an official team of observers with official equipment that's checked annually. They measure rainfall and snowfall amounts, which become the official representation, or numbers, that are used for records. They report to TV and radio networks and to news operations throughout the United States and around the world.

Are there any rules for predicting the weather?

Yes. There are quite a few technical rules, but the most important rule, one that I like to constantly remind myself of, is that Mother Nature will always humble you. Just when you think you've got the weather nailed down, Mother Nature will throw you a curveball—a shift of wind direction, a change in temperature—and you find yourself with a forecast that's completely wrong.

The main technical rule I like to follow is: always

know where the weather patterns are coming from and going to.

What tools does the weatherman need to predict the weather?

Well, the professional has quite a lot of advanced equipment at his disposal, from weather satellites and radar to computers and balloons. No single system of measurement can give the meteorologist the whole picture, so information is gathered from many sources.

How do weather balloons and radar help predict the weather?

Weather balloons are launched every 12 hours from over 200 cities around the country. They contain instruments which send back information to weather stations. As they rise, these instruments record air temperature, humidity levels, air pressure, and winds high in the atmosphere. These data help meteorologists determine how weather patterns are moving and can indicate changes that may occur, such as where weather is coming from or where it's moving toward.

Radar does the same thing. Radar works by radio waves being sent into the atmosphere and returning after bouncing off something. Radar can even sense the presence of rain and transmits this information to us along with a radar picture of the earth, or superimposes this information over a map which is what you've probably seen on television. Those colored blobs are radar pictures of cloud cover and precipitation, that is, moisture of some form—rain or snow, for example. Each color represents an increasing intensity of precipitation, with red being the most intense and blue the least. Radar can also show the speed at which precipitation—rain, snow, sleet—is moving and indicate how and where it may go. Radar is also excellent in predicting the *onset* of precipitation. Again, the colors show the amount of moisture in the air; the more intense the color, the more likely it is that there will be precipitation.

How accurate are weather-predicting computers?

Weather-predicting computers are really extraordinary. Some computers may hold data for as many as 350,000 mathematical formulas. By inputting certain information to a computer, such as temperature, type of cloud cover, and wind direction, the computer can predict—almost to the tenth of an inch—the amount of rain or snow that will fall. For example, in New York City, we had snow March 19, 1992, the day before the first day of spring. The National Weather Service computer, late the night before, had calculated 6.2 inches of melted precipitation—which is the equivalent of 6 inches of snow—and we got 6 inches of snow.

Weather-predicting computers aren't perfect, however. Even though they do contain all that data, computers are still only making *predictions*. It's important for any meteorologist to know the weaknesses of his computer and to sometimes follow his instincts rather than what the computer says. This is one area that separates the good forecaster from the bad.

How do satellites in outer space help forecast the weather?

Satellites are an absolute must for weather forecasting today. Think of them as space cameras—cameras in the sky that take pictures of, and report to weather stations around the world (including my office!) every 4 minutes. Since the 1950s, we've been able to use satellite pictures as an "early-warning system" for big storms and also as an indicator of good weather. In addition to giving us pictures of cloud cover and weather movement, satellites provide a huge amount of information the meteorologist can pass along to the public. If the satellite breaks down, the meteorologist must rely on older, simpler technology.

What are some of the simpler tools a weatherman may use?

In addition to his own common sense, a weatherman uses many pieces of equipment you may already have in your own home.

There's a thermometer, which is used to measure air temperature. Usually we're checking for the current air temperature as well as recording the high and low temperatures for each day. It's important when you install a thermometer that it be out of direct sunlight, as direct exposure to the sun can cause the temperature to record much higher. Think about what it's like on a hot day. It always seems hotter in the sun than when you stand in the shade of a tree.

Believe it or not, the thermometer was invented 400 years ago—in 1592 by Galileo.

A barometer is another instrument that many amateur weather-watchers own. A barometer is probably the most important and most useful instrument to have if you want to make your own forecasts as well as to record the weather. A barometer tells you

the air pressure—that is, how heavily the air is pressing down on the earth. Cold air has more pressure than warm air, so an increase in pressure usually indicates the approach of weather that's cooler and drier—drier because cold air holds less moisture than warm air. The barometer was invented by one of Galileo's students, Evangelista Torricelli, in 1643.

You can measure the wind speed with an anemometer, and wind direction with a wind vane. The device used to measure humidity is called a hygrometer. A hygrometer has two thermometers, one which rests in distilled water and the other which is kept dry. The difference in temperature between them indicates humidity on a scale provided by the manufacturer of the device.

What are warm and cold fronts?

Think of cold and warm air as the cause or stimulus for all weather. There'd be no wind without contrasting temperatures of warm and cold. There would be no clouds without contrasting temperatures of warm and cold. Warm air rises, cold air sinks, and in that mix all weather is created, from partly cloudy to cloudy, to rain and snow. When there are extreme clashes of cold and warm, you get violent weather. So when you see a cold front and a warm front on a map, that's the leading or beginning edge of change. Cold front—the leading edge

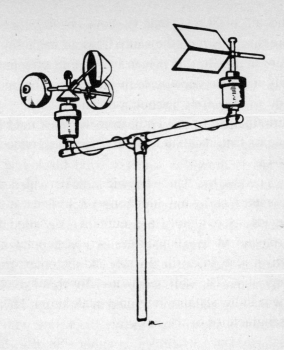

of cold air—and warm front—the leading edge of warm air—and where they meet is where weather, usually in terms of precipitation, occurs.

How do you read a weather map?

Reading a weather map is really a question of understanding the symbols. When you look at the map, whether on television or in a newspaper, lines, letters, and numbers indicate fronts, temperature, and sometimes the wind direction and speed, as well as barometric pressure. Barometric pressure is represented by solid, curved black lines. High-pressure weather systems are usually marked with an "H" and are drawn as if they are sweeping up across the map. Low-pressure systems are marked by an "L" and are drawn as sweeping down.

Fronts can be shown in three ways. A cold front, meaning the leading edge of a large, moving mass of air, is drawn as a heavy solid black line with small triangles. The triangles indicate which direction the front is moving. Like a cold front, a warm front is shown by a heavy black line, too, but is indicated by small half-circles which point in the direction the front is moving. A stationary front is a front that is, well, stationary. By that I mean it's not moving and hasn't moved in 48 hours. This type of front is shown by a heavy black line with both triangles and half-circles on either side of it. Sometimes stationary fronts can stall for days, creating little storms which travel up and down this "boundary," looking for a way to move on, or escape and continue their trips across the country.

Sky conditions are shown by small circles which indicate major cities on the map. Completely filled-in circles (usually black) mean overcast skies; half-filled circles indicate partly cloudy conditions; and empty circles mean clear skies. If rain or snow is predicted, the circle may have an "R" or an "S" inside.

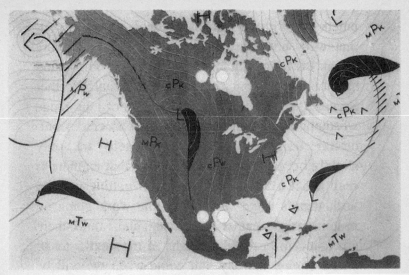

THE COMPLETED WEATHER MAP

A weather map showing several large high pressure areas across our continent.
(Neg. No. 299746. Courtesy Department of Library Services, American Museum of Natural History)

How do you make long-range weather forecasts?

The long-range weather forecast is usually supplied by a computer, and it's the 5-day forecast you see on television. If it's Monday, you'll get a weather forecast that'll cover right through to Saturday. The long-range forecast—the 5-day forecast—is pretty tricky for weather-people. There's an accuracy of about 90 to 95 percent on the first day, about 88 percent on the second day, and then the accuracy tends to drop off on days three, four, and five. Computers simply don't handle extended forecasts all that well, because the longer-term data has more variables. Today, there are even 30-day, 60-day and 90-day forecats put out by the National Weather Service. We're in the beginning stage of long-range weather forecasting, but so far the chances of predicting long-term weather accurately are only slightly better than a toss of a coin, that is, there is only about a 50 percent chance of getting it right. Still, there's some very interesting work going on in long-range weather forecasting, and I think there are going to be breakthroughs in the next 5 to 10 years. I think the day will come when we will be able to determine what the weather will be like far in advance with some kind of accuracy.

How does the ocean temperature affect weather on land?

Keep in mind that the ocean covers 80 percent of the earth, and the water in different parts of the world has different temperatures. In the winter, in our part of the planet, water cools down. Even though the land temperature may be mild—for example, in the 60s—an ocean breeze blowing across the water with a temperature in the 30s will cause temperatures on land to drop into the 40s. Computers cannot sense ocean breezes, but weather forecasters know enough to predict temperatures slightly for coastal areas. If you live in a part of the country that's near the beach, you may hear your weather forecaster predict temperatures as much as 20 degrees lower at the beach than farther inland. Sometimes when you're visiting the beach you may find that a block or two from the water you're warm, but when you walk along the sand near the surf you find you need a jacket.

What is the jet-stream?

The jet-stream is one of the keys to understanding all our weather. Surprisingly, we didn't know about it until World War II. Airplanes were flying at higher altitudes than ever before, and pilots made a startling discovery. On certain routes, pilots found that their air speeds could be reduced by head winds (Head winds are winds that blow the opposite direction the plane is traveling, the plane hits the wind "head" on.) of up to 200 miles per hour. Scientists called these winds the "jet stream" because they went as fast as jets.

Further study has shown us that the jet streams girdle the entire earth in wave-like patterns which flow from west to east. They occur much higher than where warm and cold fronts may meet—usually 10,000 to 40,000 feet above the earth. There is an old saying that if someone sneezes in the Bay of Bengal, you catch a cold on the west coast of the United States! Simply, this river of air can really produce weather effects all around the planet.

13

What makes the winds blow?

Keep in mind that all weather conditions are caused by extremes in temperature. If you have very cold air—like 10 or 15 degrees—colliding with very warm air (for high atmosphere)—like 40 or 50 degrees—that sharp contrast in temperature from cold air sinking and warm air rising will produce wind. All wind is the result of this phenomenon, the changes in temperature. Find me a spot on the earth that has very little wind, and I'll show you a spot that has very little temperature change.

What is the wind-chill factor?

Think of the wind-chill factor as the effect blowing winds have on the temperature as you feel it on your skin. The wind can make it seem much colder than it is. For example, if the temperature is 35 degrees outside and the wind is blowing only 10 miles per hour, it will *feel* like it's 21 degrees. It will feel that cold against your skin, but because the temperature is really 35 degrees, water won't freeze like it would if it were really 21 degrees.

What is a sea breeze?

During the middle of the day, warm air rises off the sands by the ocean. As this warm air rises, it moves across the open waters of the ocean and cool air begins to sink underneath it. The cool air is squeezed toward land and cools down the shore

14

Wind Chill Table

MPH	35	30	25	20	15	10	5	0	-5	-10	-15	-20
calm	35	30	25	20	15	10	5	0	-5	-10	-15	-20
5	33	27	21	16	12	7	1	-6	-11	-15	-20	-26
10	21	16	9	2	-2	-9	-15	-22	-27	-31	-38	-45
15	16	11	1	-6	-11	-18	-25	-33	-40	-45	-51	-60
20	13	3	-4	-9	-17	-24	-32	-40	-46	-52	-60	-68
25	7	0	-7	-15	-22	-29	-37	-45	-52	-58	-67	-75
30	5	-2	-11	-18	-26	-33	-41	-49	-56	-63	-70	-78
35	3	-4	-13	-20	-27	-35	-43	-52	-60	-67	-72	-83
40	1	-4	-15	-22	-29	-36	-45	-54	-62	-69	-76	-87

Wind speeds greater than 40 miles per hour have little additional chilling effect.

very rapidly. That sea breeze—moving at 10 to 20 miles per hour—will knock down air temperatures as far inland as 5 to 10 miles. People who live as far as 10 miles from the ocean will benefit from the cooling effects of the sea breeze during a hot summer afternoon.

What is a land breeze?

A land breeze is just the opposite of a sea breeze, and usually occurs in the evening as temperatures begin to drop. Air over land begins to cool and sinks back to the earth, squeezing the warm air back over the ocean, away from land.

How important is the sun for our weather?

Without the sun, there would be no weather. There would be no temperature (actually there would but it would be incredibly cold), no winds, no moisture. In fact, without the sun there would be no life. Light and heat from the sun are the energy sources which cause all weather to happen. Sunshine or rain— every type of weather happens because the heat of the sun keeps the air in the atmosphere in motion. Everything about the way our sun heats our planet is perfect—it's one of those miracles of nature. It is thought that if the sun's heat dropped by only 15 percent, our planet would be covered with a layer of ice 1 mile thick.

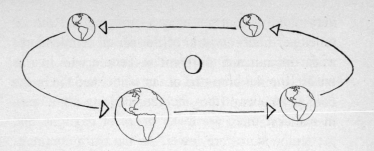

So what exactly *is* the sun?

The sun is a huge burning ball of hydrogen gas. It creates the energy that radiates across the solar system—93 million miles, to be exact—to heat our earth. Imagine traveling all that way and still being able to heat our planet. Now, that's hot!

If the earth didn't turn on its axis, what would the weather be like?

If the earth didn't spin there wouldn't be much weather. One side of our planet would be very, very hot, and the other side of the planet would be very, very cold. The way it works is this: the earth tilts on its axis at 23.5 degrees. That 23.5-degree lean remains the same as the earth rotates daily on its axis on its annual revolution around the sun. (It takes 23 hours, 56 minutes, and 4 seconds for the earth to rotate daily. The earth moves 18.5 miles per second. This means that by the time you finish this paragraph, the earth will have moved 830

17

miles.) As the earth moves around the sun during the year, different parts of the planet are tilted toward the sun and different seasons occur. In our winter, the northern part of our planet and the North Pole point away from the sun; the sun seems low in our sky, there are fewer hours of daylight, and the temperatures are lower. As the earth continues its trip, the North Pole points back toward the sun, the sun appears higher in the sky, and the days get longer.

Why can you look at the sun sometimes and not at other times?

Interestingly enough, you can look at the sun when the air quality is poor (when there are many pollutants in the air) because dirty air affects the ability of harmful ultraviolet rays to pass through the atmosphere. Pollutants give a kind of milky quality to the air in the noonday hours in Los Angeles, for example, or the industrial Midwest, or New York City. It's also possible that you can look at the sun on a winter's day because the sun, though closest to the earth during the winter, has it's rays at the most indirect angle to the earth. During the summer, when it's the most difficult to look at the sun, the sun is farthest from the earth, but the rays are more directly toward it. So, during the heart of the winter or during days of high pollution alert or an ozone warning, it is quite easy to look at the sun. It is most difficult to look at the sun during the fall, spring, or especially the summer, as the rays become more direct toward the earth.

Why is the sun different colors at different times of the day?

The color of the sun is caused by the angle at which the sun's rays pass through the atmosphere of our planet. The sun's longest, most penetrating rays are the red- and orange-colored ones. These are the strongest rays, which can penetrate the atmosphere as the earth rotates away from the sun. The red and orange rays are also reflected by high, thin cirrus clouds at around 30,000 feet. Not only do these clouds make the sun look like an orange fireball, they can cause some of those spectacular fiery sunsets.

The shorter rays of the sun—the blues, indigos, and violets—are usually broken up because of their shorter wavelengths. They are not able to travel great distances and can't make it through the clouds. For these reasons, we rarely see these colors.

The sun we see throughout most of the day is yellow, almost white. White in light is all colored rays mixed together.

How does the moon affect the weather?

Some meteorologists believe that the moon affects the weather by how much wave action will come ashore during a storm or hurricane. The phases of the moon cause the tides to change each month, too.

The moon may also have a significant effect on long-range forecasting. Many Japanese meteorologists are following the movements of the stars, the planets, and the moon. These scientists believe that the gravity of these stellar bodies have an effect on the gravitational pull of the sun, which controls weather. This is a gray area in weather research. It's something that needs to be investigated more fully. There are some indications that the moon certainly affects high tide, low tide, and wave action during the height of a storm. That's why you will often hear the term ''astronomical high tide,'' which means the moon and planets may affect the tides.

How does outer space affect the weather?

Outer space does not affect the weather. Our weather is contained within our atmosphere. Most meteorologists believe outer space has no effect when it comes to the weather. There are, however, some meteorologists who believe that outer space *does* affect the weather. They're called astro-meteorologists, and they theorize that the stars, sun, moon, and other planets do have an effect. Although there's no concrete evidence now which shows that outer space does influence our weather, it's in this area that a major breakthrough in long-range forecasting may occur.

What about sunspots?

Sunspots are "imperfections" on the sun's surface and are thought to be related to the sun's turning on its axis. Sunspots are really flare-ups—flames leaping from the surface—from the sun. And they do have an effect on our weather and climate. When there are more sunspots, temperatures can be higher, we have longer warm seasons, and the aurora borealis (see the next question for an explanation of this phenomenon) is seen more often. Some people even believe that with more sunspot activity, there will be more human activity.

What is the aurora borealis?

The aurora borealis, also known as the northern lights, is caused by light reflecting off electrical particles in the sky. They appear as streamers of light, and can vary in color from blue and yellow to bright pink and red. Meteorologists don't believe the aurora borealis has any effect on the weather, because it can be seen, in some regions such as the Arctic, nearly every night regardless of the weather. But, like everything else, this too is being studied to see if it does have an effect on our weather.

How do volcanoes change the color of the sunset?

Sunsets are directly affected by volcanic reaction. When Mt. Pinatubo in the Phillipines erupted in 1991, ash was shot some 60,000 to 80,000 feet into the atmosphere. Because the ash went so high, it will stay there for a very long time, because it will remain born by the winds aloft. Many of the shorter rays of the sun—the blue and violet ones—cannot pass through the ash and the atmosphere, while the red and orange rays will be able to. This should produce beautiful, vivid sunsets across the earth through 1994. Some people theorize that because a

This erupting volcano's ash will spew high into the atmosphere and may cause some spectacular sunsets around the world.
(Neg. No. 330411. Photo: Martin Lowenfish. Courtesy Department of Library Services, American Museum of Natural History)

lot of the rays of the sun have been blocked, it's conceivable that temperatures across the planet may decrease from three-tenths of one degree to perhaps as much as 2 degrees. This doesn't seem like much, but it could cause a marked difference in some climates and growing seasons. But that remains to be seen.

What is El Niño?

El Niño is a term coined by meteorologists to explain the unusual—some would say freak—warm waters that are appearing across the Pacific Ocean. These pools of warm water have a dramatic effect on the weather across the U.S.A. El Niño causes mild masses of air which can stretch all the way across western Canada and Alaska, producing mild winters and melting the snow cover in the northern parts of Canada and Alaska. El Niño was first identified in 1925, but only recently have scientists felt that it is producing some dramatic storms on the west coast of the U.S. Although El Niño has only been identified as a weather cause for about 10 years, weather history indicates that this is a phenomenon that has been occurring for a long time.

What weather conditions cause the most delays in airline flights?

There are a couple of conditions that can cause delays, depending on the airport. Fog, bad visibility from rain or snow, low ceiling (low-hanging clouds), icy runways, and snow are the biggest weather factors at airports. Although most airplanes are equipped with radar, sometimes the weather is so bad that it can affect the air traffic on the ground. Some airports—such as Dulles International Airport outside Washington, D.C., LaGuardia Airport in New York City, the Chicago O'Hare Airport, and Denver's Stapleton Airport—experience tremendous crowding on a day-to-day basis, and even a short 15-minute downpour can cause a back-up of planes for hours.

It's been said that wars have been won or lost according to a weather forecast. Is this statement really true?

Well, the weather *has* been very important to commanders trying to get an edge in a war. For example, during the Korean War, a Korean invasion from the north in 1950 happened during the monsoon season. There's no question that the North Koreans were strong on the ground and the Allied forces were big in the air. By invading during the heavy-rain season, the North Koreans helped diminish the Allies' air-support strength. Advantage: North Koreans. Knowledge: weather.

1939. The Germans attacked the Poles, and they did it with their famous tanks, the panzers. The Germans were well aware that the best time of year to

WEATHER WORDS

Blizzard As a word to describe winter weather, "blizzard" first appeared in an 1870 newspaper account in the Esterville, Iowa, *Northern Vindicator*. It comes from the German phrase *Der Sturm kommt blitzartig*, which means "the storm comes lightning-like."

Cyclone "Cyclone" was first used to describe the weather in 1844 by a sea captain. It means "coil of a snake."

Hurricane The word derives from the Caribbean word *huracan*, which means "great wind."

Meteorology The root of the word "meteorology" is "meteor." Thousands of years ago, meteors were thought to affect the weather. "Meteorology" most likely comes from the title of a book by the ancient Greek philosopher Aristotle which was titled *Meteorologica*.

Mistral Mistral means "master" in French. It's used to describe powerful north winds.

Smog The word was first used in 1905. It did not find its way into the English language until the middle of this century, when a reporter mixed up the letters for "smoke" and "fog." His editor created the word "smog" for a headline.

Weather "Weather" comes from an ancient European language more than 5,000 years old. Its roots are *we* which means "wind," and *vydra*, which means "storm."

Did you know that the Eskimos have more than 30 words for snow?

invade by land is when it's driest, and September is the dry season in Poland. Advantage: Germans. Knowledge: weather.

1944. Normandy, France. This was a massive operation, and British and American meteorologists needed a very refined forecast to assist in the Allied decision-making process concerning the invasion of Normandy. The period chosen was early to mid-June of 1944. The conditions needed to be perfect. No heavy winds to hinder the Navy in the English Channel. Visibility needed to be at least 5 miles for the bomber pilots. Advantage: Allied armies. Knowledge: weather.

2

Wet and Wild: From Clouds to Downpours

What is a cloud made of?
All clouds, whether they're patchy fog or high, thin clouds at 30,000 feet, are made of condensed moisture. Most times, with warm air rising up and over heavier cold air, clouds will form. But they will form in many variations, and will look different and will produce many different kinds of weather phenomena.

How many different kinds of clouds are there?
There are many varieties, which take many forms, but they basically fall into three categories.

The first is cumulus. Its name comes from the Latin word meaning "to accumulate or heap upon." So cumulus clouds are clouds that accumulate one

Cumulous clouds. Notice how the clouds seem to be piling up on top of each other or *accumulating*.
(Neg. No. 6358 fr. 6. Photo: R.E. Logan. Courtesy Department of Library Services, American Museum of Natural History)

on top of another. They are puffy and look a lot like piles of cotton balls.

The second kind of clouds is stratus clouds. Stratus clouds usually have layers to them; the word comes from the Latin word *stratum*, which means sheets or layers. They're dull, usually gray clouds that hang low over the ground. Sometimes they hang so low they are actually fog.

28

Stratus clouds. These clouds are like a layer of clouds across the sky. Often they are snow clouds.
(Neg. No. 6356 fr. 8. Photo: R.E. Logan. Courtesy Department of Library Services, American Museum of Natural History)

The third type of cloud is cirrus clouds. These are the wispy clouds that are so high up in the atmosphere that they're made entirely of ice crystals. Cirrus clouds can be blown into wisps by the jet stream. Some people call these clouds "mare's tails."

All of the other clouds you see are made up of combinations of these three types of clouds. For a

29

Cirrus clouds. These are the wispy clouds high in the sky. Here they are shown reflected against the sunset.
(Neg. No. 6358 fr. 9. Photo: R.E. Logan. Courtesy Department of Library Services, American Museum of Natural History)

long time people thought there were so many types of clouds that it would be impossible to classify them. In 1803 an amateur weatherman, Luke Howard (who was also an English pharmacist) identified 10 distinct categories based on the three basic cloud types. His system was so easy that meteorologists still use it today.

What kind of clouds bring what kind of weather?

In the summertime you may see clouds stacking up, building what looks like towering mountains. Sometimes they appear gray or black. In the Midwest, sometimes they'll appear purple or green. These clouds are cumulonimbus clouds. They produce thunder and lightning and are probably the most dangerous type of cloud—dangerous if you're caught in a storm, that is.

Rain clouds are usually low, gray clouds. They can be nimbostratus, stratocumulous, or just plain stratus clouds.

Altonimbus clouds are the kind that appear on partly cloudy days. They are big and puffy and can block the sunlight at times, but generally do not cause any kind of precipitation.

What is relative humidity?

Well, have you ever gone outside and had your hair frizz up? Or have you been outside and started to perspire when you hadn't even exercised? How about those glasses that get so foggy that you can't see? These conditions are all caused by relative humidity—that is, the amount of water vapor that's in the air.

Relative humidity is very important to the meteorologist because it is a measure of how much moisture is in the air. By measuring moisture content, the meteorologist is able to figure out the amount of rain or snow an area is likely to get.

Why does it rain even when the humidity is not 100 percent?

Humidity does not have to be 100 percent for it to rain. Many people think that it has to be, but it doesn't. If the humidity is 65 or 70 percent there could be dry spots in the atmosphere, but rain is still possible.

Where does dew come from?

Dew is a phenomenon that usually occurs in the late spring or early autumn. What happens is that the ground stores up heat during the day. During the evening, the air, if calm enough, will begin to rise and cool. With calm conditions, when the cool air hits the very warm air rising from the ground, you get condensation. This usually shows up on blades of grass, flowers, or metal objects like cars. Dew does not come from clouds. It is simply the exchange, at ground level, of warm rising air and a very cool layer of air at the surface forming droplets of moisture on the surface.

What is frost?

Frost is the same thing as dew. The only difference is that when air condenses on cold surfaces at night, it forms ice crystals and simply freezes. If the exchange of very warm air rising off the ground and cool air at the surface does not hit 32 degrees, you get frost.

What causes thunderstorms?

Thunderstorms are the result of very warm air rising rapidly in the atmosphere (up to 6,000 to 40,000 feet) and very cold air driving down (from over 40,000 to 60,000 feet). Usually, they are in advance of cold fronts and are the result of these extreme combinations of very warm and very cold air banging into each other. Thunderstorms produce lightning, thunder, and a good deal of wind.

Where does lightning come from?

This is a complicated process that takes place in layers of the atmosphere from 6,000 to 55,000 feet. The tremendous exchange of warm air and cold air produces friction. The friction causes separations of ions (positive and negative electrical charges in the clouds). In essence, what we see as lightning is the friction of very warm and very cold air rubbing together. The sound that we hear resulting from that friction is thunder. If you are near a lightning strike (which you don't want to be) you might hear a *vitt* or *zitt* sound. This is the real sound lightning makes, not the crash of thunder.

Are there different kinds of lightning?

Yes. Lightning appears in many different shapes and forms. One form of lightning is called streak lightning. Some people call this fork lightning, as it looks a bit like a giant is throwing forks down to earth. It's the most commonly seen lightning, and usually occurs within a cloud.

Heat lightning is usually the kind you see at a distance on a warm summer night. You do not hear thunder with this kind of lightning.

Sheet lightning is what you see as a whitish glow in different layers of the atmosphere.

Cloud-to-ground lightning, which is a kind of streak lightning, is the most dangerous kind. Streak lightning causes the most damage, in the form of injuries in open and rural areas, rather than in cities. Tall buildings in urban areas are good conductors of electricity for cloud-to-ground lightning, and these buildings are grounded to prevent electricity from causing damage.

Interestingly enough, more men are killed by lightning than women; they are, on the average, outdoors more often during thunderstorms.

It's said that lightning doesn't strike twice. Is this true?

No. This is one of the countless folktales about weather that's not true. For example, in New York City, scientists observed on the World Trade Towers and the Empire State Building anywhere from 22 to 36 successive strikes of lightning on both buildings during a given storm. These strikes were observed within periods as short as 10 to 15 seconds.

What color is lightning?

Lightning is usually white. Occasionally, it might look yellowish if it's reflected against artificial lighting. It can also look bluish, or blue-toned. When air is very, very moist, it may have a reddish hue because of the charged hydrogen.

There are many unproven observations about lightning. Forest rangers say white lightning may cause more forest fires; bluish lightning is associated with hail, but again, these are unproven observations.

Why do you see lightning before you hear thunder?

Lightning travels 2,000 feet in roughly one-hundredth of a second. Thunder travels roughly 1,100 feet per second. You can certainly see it before you hear it.

You can tell how far away a streak of lightning is by counting the seconds between the flash and the thunder. Divide the number of seconds by five and you'll know the distance from the lightning in miles.

When you see a flash of lightning, start counting

immediately. One thousand one; One thousand two; One thousand three. When you hear the first rumble of thunder, stop. If you've counted to ten, divide ten by five and you know the storm is two miles away.

What time of year do most thunderstorms occur?

It varies. Thunderstorms can happen at almost any time, but they usually occur in late spring and summer. Most thunderstorms, especially in the tropics, happen in the afternoon after the sun has heated up and stirred up the air. Inland, thunderstorms happen after long stretches of warm weather.

Any part of the country is susceptible to thunderstorms but the southeast and the Midwest have the most violent thunderstorms. In the most violent of thunderstorms, tornado activity can sometimes occur.

Where is the safest place to be during a thunderstorm?

The safest place to be is in a car. The rubber in the tires will act as a conductor or travel path from the metal of the car to the rubber and to the ground, taking the electric charge away from you.

If you're caught outdoors during a thunderstorm or thunderstorm activity, the best thing to do is to curl up into a little ball flat on the ground. Avoid trees, water towers, or any kind of high, exposed rod, and stay away from any kind of peak or ridges. Lightning searches out the highest thing around. Get out of the water, get away from open beaches, stay away from golf courses, stop all outdoor games, don't ride a horse, and, of course, don't ride a bicycle or drive any machinery like a tractor during thunderstorm activity. If possible, hurry indoors as quick as you can. If you can't get indoors, just stay in the open. You'll get a good soaking, but it's safer to be wet than under a tall structure or tree.

The chance of being struck by lightning is very slim. In fact, you have a bigger chance of sitting next to someone on the bus with the same first name as you. But lightning is nothing to scoff at, no matter how small the risk—you should make safety your first priority.

A baseball team is out on the field and a thunderstorm breaks loose. Why is the pitcher the most likely one to be hit by lightning?

Lightning will strike the highest point, and the highest point on a playing field will be the mound, and the mound would be where the pitcher is.

How do you make rain?

Over the years at Channel 2, I've had people call me up, announcing themselves as rainmakers, "raintakers," "rainfakers," and so on. But it's impossible to take blue skies and fair-weather clouds and make it rain. It's a complicated process, actually, beginning in clouds that are capable of producing rain. Rain has to contain the proper amount of water droplets. Each raindrop is a collection of tiny water molecules clustering around a particle of dust. Raindrops will form only if there is enough dust and smoke in the air. These particles are called "condensation nuclei." If the air is clear, with no pollutants or dust, there are not enough of these particles and clouds won't form.

Why do some clouds look black?

Some clouds look dark and gray because they are full of water droplets and no sunlight passes through them. The heaviest rains come from the darkest clouds. The lower, lighter-gray clouds, usually nimbostratus clouds, produce slow, steady rains—the kind that can last for days.

What's the difference between a shower and a cloudburst?

Weather forecasters have descriptions for differing amounts of rainfall. If less than 1/48 of an inch falls in an hour, this is considered light rainfall, and heavy if more than 1/6 of an inch falls in an hour. A cloudburst is caused by strong updraft and winds within a cloud. These violent downpours usually form along cold fronts.

How is hail formed?

Hail is the result of violent wind surges within thunderstorms, and is usually found only within thunderstorm clouds. Very violent surges of updrafts cause water to rise within clouds and into colder air and then freeze. These frozen particles drop, pick up more water condensation, and are thrown back up into the colder air. The frozen raindrops begin to take on a layered look, until finally they get caught in a downdraft which forces the drops to come down at great speeds. These drops become so heavy that they fall to the earth as hailstones, so rapidly that they don't have a chance to melt. Hail can be the size of golf balls or even oranges. The largest hailstone ever recorded was found in Coffeyville, Kansas, in 1970. It weighed 1.7 pounds. That's big enough to shatter a window or even dent a vehicle—big enough to cause serious accidents.

What is a rainbow made of? Why does it disappear so quickly?

Rainbows always have the same "ingredients." They need a presence of moisture in the atmosphere at low, middle, and high levels. Rainbows need some kind of light shining on them, and in nature, this light is from the sun. The observer should be at an angle where he or she can see the light reflecting off the water vapors that are in the air. It's the refraction of that light which produces colors—

colors that pass from the sun through the water droplets to the observer's eye. In short, rainbows are refractions of sunlight through water droplets.

What is sleet?

Sleet is made up of ordinary raindrops which, unlike hailstones, are not tossed about in the upper levels of clouds. Sleet is formed when rain travels down toward the ground and passes through very cold air. The drops then freeze into little bits of ice. So sleet is just frozen rain; weather forecasters sometimes use these terms interchangeably.

What then is snow?

Sometimes water vapor in clouds cools so that it freezes into snowflakes rather than collecting into raindrops. Snowflakes are not formed in the same manner as hail and sleet are. Snowflakes are formed directly out of water vapor in the clouds. Hail and sleet were raindrops first.

Why can it rain one side of the street and not on the other?

This very often occurs in the summertime. It is not a uniform pattern of two, three, or four hundred miles of rain or snow heading in a certain direction. It's usually one cloud passing by, loaded with just enough moisture to drop rain on a very small area.

Snow is frozen water vapor. Some people say that no two snowflakes are alike. These two pictures show just how different two snowflakes can be.
(Neg. No. 2A3018 and 2A3014. Photos: A.J. Rota. Courtesy Department of Library Services, American Museum of Natural History)

It's conceivable that you could be standing on one side of your town getting rained on, while the the other side of town is sunny. One cloud, isolated, with enough moisture, can rain on your parade.

How much rain is in a foot of snow?

The general equivalent is 1 inch of rain equals 10 inches of snow. Let it snow, let it snow, let it snow!

What part of the world receives the least amount of rainfall?

In Helwan, Egypt, the average annual amount of rainfall is 1.33 inches. Closer to home, in Greenland Ranch, California, the average is 1.56.

What part of the world receives the greatest amount of rainfall?

In Mount Waialeale, Hawaii, the average rainfall is approximately 471 inches per year. In Cherrapunji, India, the average annually is 450 inches. Once it even added up to 16 feet in 15 days. Rain, rain, go away!

What was the largest snowfall in a single day in the U.S.?

In Silver Lake, Colorado, it snowed 76 inches in a 24-hour period . . . and it happened in April! What ever happened to "April showers bring May flowers"?

3

What's Hot? What's Not? Climate and Temperature Around the World

What's the difference between climate and temperature?

The word "climate" comes from the Greek word *klima*, meaning region or zone. Climate is the history of wind, temperature, rain, snow, drought, violent storms, hurricanes, and so on, over a long period of time. It is a combination, or synthesis, of weather patterns for a given area.

Temperature is the measure of the heat-caused molecules in motion at a specific point in time. The movement of molecules causes heat, which is energy. This causes all of our weather.

Is our climate changing?

Many people believe that the climate *is* changing, due to the increase in heat known as global warming. I believe that climate is cyclical. If you measure the temperature of the United States in the last 100 years, for example, it has changed very little.

However, it is interesting to note that 1990 and 1991 were the warmest years ever in New York City history. The winter of 1991–92 was the twelfth warmest ever. In the area from Washington, D.C., to Boston—an area that is greatly developed, and one that has expanded quite rapidly—studies show less record-breaking nighttime cold temperatures than at any time in the history of recorded weather (back to 1899). This suggests that because man has altered the environment by creating a "concrete zone," or "canyon," of houses and office buildings along the east coast of the U.S., the ability of this area to lose heat during the night has consequently been affected. These changes in temperature are a new phenomenon, but much more evidence is needed in order to prove that our climate is in fact getting warmer. There is no need to panic over this.

Why do some places have four seasons and others don't?

In the United States, the winter solstice usually begins around December 21st. At this time, the sun is 91 million miles away; it is closest to the earth as it gets during its cycle, but the angle of its rays to the earth is wider and therefore weaker because the earth is tilted slightly away in the northern hemisphere. That gives us winter. The southern hemisphere gets summer at this time, because the tilt of the earth makes the sun's rays more directly focused on it.

This effect is reversed during the summer solstice. The earth is farthest from the sun (94 million miles), but the sun's rays in the northern hemisphere are more direct, so the weather is warmer. Conversely, the southern hemisphere now has an indirect angle to the sun. Therefore, it is winter there.

The passage of the seasons from the winter solstice to the summer solstice takes time, and the earth's angle changes gradually, changing the temperature, too. That's why some places around the earth have autumn and spring. Since, as we learned earlier, the earth only tilts at 23.5 degrees, some parts of our planet only receive short, indirect rays and others long, intense rays all year round like the Sahara Desert of Africa and parts of Brazil.

Why is it hot at the equator?
The equator receives longer rays all year long, and because of this there's not much change in air masses. The only two fronts that mostly occur are the equatorial air mass and the subtropical air mass, brought to the equator by means of the trade winds,

which are laden with moisture and produce a lot of rain. We call this area the doldrums.

As you learned in Chapter 1, the earth rotates around the sun and spins on its axis. The North and South poles receive the least amount of sunlight and the poles spend a large part of each year in darkness or twilight. Therefore, they receive the least amount of heat and are the coldest places on earth.

Which is colder—the North Pole or the South Pole?

One can argue that the South Pole is colder. The South Pole has a winter-time average temperature of 89 degrees below zero. On May 11, 1957, a temperature reading of 100.4 degrees below zero was recorded near the South Pole!

However, in 1938 in Oinekon, Russia, which is near the North Pole, a temperature reading of 108 degrees below zero was recorded.

It's so cold at both poles that if you were to shed a tear, the drop would freeze halfway down your cheek. Your breath would form ice crystals in mid-air; it would be like a wall in front of your nose. Gas vapors from pumping a car would not make it into the tank because they would freeze in midair.

Is it always hot in the desert?

Absolutely not. It's quite common in the deserts of the United States—in Las Vegas, Nevada, for example—that you could have daytime temperatures in the 90s and at night have temperatures drop into the 40s. The heat is extreme during the day, but under clear skies a lot of that heat escapes into the open night sky, and the temperature can drop 40 or 50 degrees in a 12-hour period.

The desert can have great swings in temperature, and also a lot of variation in the amount of precipitation it receives. It's not uncommon for a desert to go for a long period of time without rain, and then get torrential rains. Sometimes it will even snow in the desert.

Does it always get colder as you climb to the top of a mountain?

The answer is yes. For every 3,000 feet that you go up, the temperature will drop a degree or two. This is because the atmosphere is thinning and heat from the sun's warming rays can escape more easily.

Are droughts becoming more frequent, or are they cyclical?

It is my opinion that all weather is cyclical. We go through periods of many hurricanes and then few, snowy winters and snowless winters, heavy periods of rain and then drought. The question is not whether droughts are occurring more often, but whether we can figure out the cycles in which droughts occur.

47

The Farmer's Almanac: **Predictor or Pretender?**

The Farmer's Almanac tells us a lot of things, one of which is the hope that it can predict the weather way in advance. The reality is that weather can be predicted well only within a 24- to 48-hour period. The 5-day forecast, as you get to day four or five, has less credibility. The National Weather Service puts out a 30-, 60-, and 90-day weather outlook, but it would be the first to admit that long-term forecasts are basically just guesses. To predict the weather a year ahead is impossible, but there are people out there who would like to know the weather that far in advance. *The Farmer's Almanac* provides that information. Whether or not it's right depends a lot on guesswork and on looking at weather cycles.

Why do we seem to be getting less and less snow each winter in the United States?

This may not in fact be true. Statistically, all relevant information has to be gathered in order to verify such a statement. For example, in the winter of 1991–92, New York City received nine inches of snow (with the average being 26 inches). Nine inches was certainly a very little amount, but not a record. In fact, 2.3 inches of snow was the record (set back in the 1970s). However, in the same year that New York City received only nine inches of snow, Syracuse, New York, had its snowiest winter ever, with 152 inches of snow. We seem to be in a

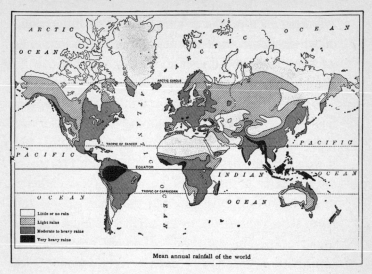

Mean annual rainfall of the world

Weather across the earth is caused by the heat of the sun and by the movement of the winds. Notice that the areas on the planet that receive most of the heaviest annual rainfall are in the tropics where the daily weather remains constant throughout the year.
(Neg. No. 269856. Courtesy Department of Library Services, American Museum of Natural History)

period of less snow (most notably on the east coast, in the inner mountains of the Rockies, and at the ski resorts of New England), but it must also be noted that winter weather, like all weather, is cyclical. We should not be too quick to blame apparently lessening snowfalls on global warming, the depletion of the ozone layer, or pollutant buildup trapping

49

warm air in the atmosphere. They could all be contributing factors, but cycles in weather may be *the* most important factor in determining who gets a lot of snow and who doesn't.

Which is the driest state in the U.S.?

Nevada averages the least amount of rainfall, with only 8.8 inches per year.

**What is the hottest temperature ever re-
corded?**

In 1922 in Libya the temperature reached
136.4 degrees Fahrenheit. In the United
States, in Death Valley, California, the temperature rose to 134 degrees Fahrenheit in
1913. Anyone for a tall, ice-cold glass of
lemonade?

Where is the hottest place on earth?

In Dallol, Ethiopia, the average temperature for the year is 94 degrees Fahrenheit.

Where is the coldest place on earth?

In Vostok, Antarctica, the average temperature for the year is − 72 degrees Fahrenheit. *Brrrrrrrrr!*

**Is there a permanent sheet of ice somewhere
on the earth?**

In the Arctic and Antarctic, where it is always
cold, the ice is up to 985 feet thick. This
must be an ice-skater's dream vacation
spot!

4

Stormy Weather: Hurricanes, Tornadoes, and Other Big Ones

What is a hurricane?

A hurricane is a storm which develops off the coast of Africa near the equator and has characteristics that make it more potentially dangerous than other types of storms. It has an eye, which means winds blow in a circle around a calm spot about 15 miles across. Hurricanes have sustained winds of at least 74 miles per hour. A single hurricane can have a life of up to a couple of weeks as it swirls across Africa into the Caribbean, the Gulf of Mexico, and the east coast of the United States.

The hurricane season begins in June and lasts into November. The chief period for hurricanes is usually late August and September. They are marked by torrential rains, a storm surge (a wall of water which moves just in front of the eye of the storm). It is this wall of water that causes the most damage and deaths, as it can rise 10 or 20 feet. Rains can be up to 10 to 15 inches over a fairly short period of time.

The origin of the word "hurricane" is not quite clear. It appears to have come from the Caribbean term *huracan*, meaning "great wind." It is interesting to note that typhoons, which occur in Asia, are the exact same kind of storms as hurricanes. The word "typhoon" comes from the combination of the Chinese words *ty* and *phung*; *ty-phung* also means "great wind." Other names for hurricanes are willy-willies, *baguios*, and cyclones—but these are all referring to the same phenomenon.

What makes hurricanes dangerous?

Hurricanes are dangerous in a number of ways. It's impossible to do any sailing or navigating through a hurricane.

If you see a double flag with a black square in its middle it's a hurricane warning. That means that all ships should head to port, and should even be taken out of the water if possible.

Second, the storm surge of a hurricane (as described in the previous question) can cause massive beach erosion. A strong storm or tidal surge can wipe out 600 to 700 feet of oceanfront. With the tremendous development of homes along the east coast of the United States, a major hurricane (which we have not seen hit the east coast in 40 or 50 years) could cause billions of dollars' worth of damage. Hurricane Hugo, which hit Charleston, South Carolina in 1989, *did* cause over a billion dollars' worth of damage, and that was a storm that hit one area. If a hurricane skirted the east coast of the United States, and came all the way up to portions of Virginia, New Jersey, and New York and Long Island, the damage could be even greater.

Hurricane winds, which can gust upwards of 220 miles per hour, can cause structural damage—planes, objects such as glass, uprooted trees, and garbage cans become missiles and fly through the air. The most dangerous part of the hurricane is the sustained winds around the eye of the storm, which can cause structural damage. Many people are fooled by the eye of the storm, thinking the storm has passed. But the storm is only half over, and savage winds will return.

The front of the hurricane is usually more violent than the part that follows the eye. Many people believe the eye (the calm before the storm) means the worst is yet to come, but this is not quite true. What happens is that many people think the storm is over with and they come out. The reality is that it is not; people are outside and exposed, the storm resumes, and that's when injuries and death happen. So it is very important to follow a hurricane's progress. Listen carefully to newscasts about where the eye is moving, because the eye is an indication that more of this storm is going to hit.

What is the eye of the storm?

The eye of the storm many times shows up on satellite as a pinhole. It is the direct center of the storm, and can be anywhere from 6 to 22 miles in length. It's usually an area of calm—blue skies, sunshine. Birds can fly inside the eye of a storm.

The smaller the eye, the more violent the storm. The lowest pressure of the storm, and its highest temperature, is usually located within the eye. If you were to fly into the eye of a storm (which hurricane pilots do), you would see violent storm clouds surrounding it. You'd see lightning, hear thunder and the roar of the wind in the wall of clouds around the eye. This is where you'd find the strongest winds in the hurricane.

How do forecasters measure the strength of a hurricane?

Hurricanes are measured in categories, from one to five.

Category one hurricanes have winds of 74 to 95 miles per hour, with a storm-surge wave height of 4 to 5 feet above normal. There's no real damage to buildings; the damage is primarily to mobile homes, shrubs, small trees, coastal roads, and small piers or docks.

Category two hurricanes have winds of 96 to 110 miles per hour, with storm surges of 6 to 8 feet above normal. Some roofs are damaged, doors and windows to buildings can be blown out, mobile homes and some piers could be affected, low-lying escape routes can flood, and small craft can break their moorings.

Category three hurricanes have winds of 111 to 130 miles per hour, with storm surges of 8 to 12 feet above normal. Category three causes some structural damage to small residences and mobile homes, flooding near coasts which can destroy small structures, and larger structures can be damaged by floating debris. Terrain (less than five feet above sea level) can be flooded as far as 6 miles inland.

Category four hurricanes have winds of 131 to 155 miles per hour, with storm surges of 13 to 18 feet above normal. These can cause extensive damage to roofs and small residences, major erosion of beaches, and major damage to structures near the shore; terrain less than 10 feet below sea level may be flooded, requiring massive evacuations of residential areas as far as 6 miles inland.

Category five hurricanes are the most dangerous, with winds greater than 155 miles per hour, and storm surges greater than 18 feet above normal. These storms can cause complete roof failures, damage to many residences and industrial buildings, including complete building destruction (small utility buildings can be blown away), major damage to lower floors of all structures located less than 15 feet above sea level and within 500 feet of the shoreline, massive evacuation of residential areas within 5 to 10 miles of the shorelines may be required.

What is the best thing to do in a hurricane?
Stay indoors in a structure that is steel-beamed based, and listen very carefully to all weather forecasts. It's also very important to bring everything indoors that is not tied down—furniture, bicycles, garbage cans—you should board up all windows. You should evacuate from all coastal plains and from all areas that are close to beaches when told to do so, and not take any unnecessary risks.

The most important thing to remember is that hurricanes are very, very serious business and that lives are at stake. Hurricane warnings and watches are broadcast on radio—the Coast Guard has its own weather band that you can listen to—so between television and radio you can closely follow the progress of a storm. Nowadays, there is absolutely no reason why people cannot be fully aware of hurricane activity and take the necessary precautions.

Why do we name hurricanes?

Before 1944, there were no ways of predicting or watching for hurricanes. Since then the National Weather Service, in coordination with both the Air Force and the Navy, has an annual conference, where the names to be used in hurricane identification are decided upon. The list always begins with a name beginning with the letter "A" and goes through the alphabet. Sometimes if it looks like it's going to be an active season, 12 or 13 letters may be used. The first hurricane of the season is usually identified by a girl's name, let's say Alice, and the second one would be male, let's say Bob. (I'm still waiting for mine, Hurricane "G." Or for Hurricane Irving—I like the sound of that: "Irving is moving through the Gulf of Mexico." Or "The G-man is in the Caribbean." Weatherman's fantasy.)

Just how much damage can a hurricane do?

Hurricanes can really be killers. On September 8, 1900, in Galveston, Texas, over 6,000 people lost their lives when a wall of water hit the town and caused tremendous amounts of flooding. The water swept through the city, and many people drowned.

In 1938, a hurricane rearranged how Long Island, New York, looked. Long Island used to be shaped like a huge sneaker. When the hurricane of '38 struck, the wall of water, the surge, drove from the point of Long Island inland, creating what is now the north and south forks.

How does a hurricane change from being shaped like an eel to being shaped like a perfect ball?

Hurricanes will change according to the wind flow within the storm. Sometimes the jet stream or the winds high aloft will shear, or cut away, part of the hurricane, and make it look eel-like. Sometimes the hurricane will stall, shift, and then move in a different direction, and as it's going through that transformation, it changes from its spiraling-ball shape to being eel-shaped. When a hurricane is shaped like a perfect ball, it's usually a well-defined hurricane, with gales extending out 175 to 200 miles per hour in the center, and the hurricane force winds extending out some 75 to 80 miles from the eye of the storm.

Who are the "hurricane fighters"?

Hurricane fighters are pilots who fly into the eye of the hurricane. The planes take off from Galveston to fly over the Gulf of Mexico, and from Miami to fly over the Atlantic Ocean and the Caribbean. The closer the hurricane comes to the east coast of the United States, the more frequently the flights are taken. The pilots fly right through the storm to the eye, where the planes' instruments measure the barometric pressure, the highest winds, and the wave heights outside the eye of the storm. The information is then radioed back to the Hurricane Center in Coral Gables, Florida, which then gives intragovernmental reports to all of the weather reporters around the country for distribution to the public through radio and television. These hurricane pilots are absolutely essential for pinpointing a storm's position and strength, and for predicting where it will hit land.

What is a monsoon?

The word "monsoon" comes from the Arabic word *mausim*, which means season. Monsoons are seasonal winds of the Arabian Sea and Indian Ocean that blow for 6 months from the northeast, then reverse and blow for 6 months from the southwest. Monsoons are quite common in India and southern Asia. They can produce mammoth rains and cause tremendous amounts of flooding.

Warm air blows inland across water and begins to rise. As it approaches land, this warm air rises to pass over the mountains and as it gets higher and higher, water vapor condenses, or forms droplets. These water-vapor droplets then become rain. Since this condition lasts about 6 months, torrential rains can occur. The monsoon season usually lasts from June to October.

During the winter, the monsoon is reversed. The winds blow from mountains to oceans. This is the dryer season.

What is the difference between a hurricane, a tropical cyclone, and a typhoon?

These are all names for the same thing. They originate in tropical areas and produce violent winds, heavy rains, and storm surges. These violent storms are called by different names in the different parts of the world where they appear—hurricanes in the Atlantic, typhoons in the Pacific, and cyclones in the Indian Ocean area.

What is an anticyclone?

An anticyclone is just about what it sounds like. When meteorologists use words like "cyclone" and "anticyclone," they don't mean violent storms such as tornadoes, which are sometimes mistakenly called cyclones.

Cyclones and anticyclones are wind systems that are created when the winds high aloft in the atmosphere meet; they are mirror images of each other. Cyclonic wind systems spin counterclockwise (in the northern hemisphere) around a low-pressure system. Anticyclones rotate clockwise and spin around a high-pressure system. One trait cyclone and anticyclones have in common is that they can cover thousands of square miles. Anticyclones are usually shown on the weather map as highs and usually bring blue skies.

What is a tornado?

The word "tornado" comes from the Spanish word *tornado* which means thunderstorm. I regard tornadoes as the most violent of all storms. A tornado is a swirling mass where very cold and warm air meet and form a funnel-shaped cloud in which winds blow at an awesome speed directly upwards. In that upward-funnel cloud, the speeds can go as high as 250 to 300 miles per hour.

These violent storms usually are only about 250 yards wide in their middle and last for only a very short period of time. They may only touch down on earth for 100 feet but can travel for hundreds of miles. Most occur between three o'clock in the afternoon and eight o'clock at night because the air has warmed throughout the day and become unstable. Tornadoes travel at 20 to 40 miles per hour, sometimes as fast as 50 miles per hour. It is when they move slowly, sometimes as slow as 5 miles per hour, that they cause the most damage.

The Midwestern United States experiences more tornadoes than any other part of the world (although tornadoes can strike anywhere). As many as 500 to 600 can hit each tornado season in the Midwest.

How fast are the winds inside a tornado?

Their speeds will vary, anywhere from 125 miles per hour to 300 or more miles per hour. It's very difficult to measure the upward spiraling motion, because nothing can survive being inside of a funnel cloud. As a matter of fact, the energy is so intense and severe at the base of the funnel that sometimes when one later follows the path of a tornado, burn marks can be found on the ground where the tornado touched down. Those people who have seen tornadoes have said the noise is deafening, like the roar of an airplane.

Where do most tornadoes occur?

Generally, most tornadoes occur in Kansas, Texas, Oklahoma, and Iowa. The Midwest is a perfect zone where warm and cold air can collide. Cold air comes from Canada and warm air travels north from the Gulf of Mexico.

What is the best thing to do if you're caught in a tornado?

If time permits, go inside a steel-framed structure. If there is a basement or cellar in the structure, head there. The southwest corner of the basement is the best place to sit, as most tornadoes travel from the southwest to the northeast. It's always a good idea to keep away from gas lines.

Stay away from all windows. (If you are in a

63

house, stay in the basement—in the southwest corner.) If you are in an office building, stay on a lower floor.

If you are caught in an open field, it is best to lie at a right angle to the storm, since tornadoes tend to move to the north and east. So you would lie at an angle to the south and west from the tornado. If possible, lie flat in a depression or a ravine. And be sure to keep a watchful eye to the sky.

What is a tornado called when it appears over water?

It's called a waterspout. Instead of debris lifting up through the swirl, producing that dark cloud of dust a land tornado would have, a waterspout produces a funnel of spray. A waterspout—going from water to land—maintains its tornadic activity. Likewise, a tornado—which appears on land and then goes over water—usually keeps its same characteristics as well.

Why aren't giant tornadoes common in the northeast, northwest, or southwest U.S.?

Coastal ocean waters in these areas cool the land down enough so that there's not that awesome violent collision of warm air from the gulf of Mexico and cold air out of Canada. The ocean waters cool the tornado before it gets started.

What is the worst tornado on record?

It's said that the tornado which occurred on March 18, 1965, was the most devastating. It moved from Missouri through Illinois and finally into Indiana within 31/2 hours. It killed 689 people.

What is the strongest wind ever recorded?

On April 12, 1934, a wind at Mount Washington was clocked at 231 miles per hour.

What size was the largest ocean wave ever to hit the earth?

The biggest wave ever *recorded* was measured at 112 feet high on February 7, 1933, in the Pacific Ocean.

5

Spring Fever and Other Biometeorological Facts

Is there really such a thing as spring fever?

Spring fever can be both a mental and a physical response to the arrival of spring. Being cooped up indoors, especially if the winter has been a severe one, produces what's called cabin fever. The dry heat in homes and offices takes its toll, too. People become tired or lack of sun causes people to become grumpy, too. Studies have shown that sunlight can improve a person's mental disposition. And that's the beginning of spring fever.

The physical changes may simply be the springtime adjustment of the body changing its blood circulation in reaction to the warming of the external environment. As the body begins to acclimate to the warmer environment, this feeling of renewed energy is said to be the beginning of spring fever.

Can extreme changes in weather affect your health?

Yes. Many times, if you have problems with your cardiovascular (heart-lung) system, a sudden change in the weather, especially a severe drop in temperature, means the body must generate more warmth and to do so the heart must beat a little faster to warm the body. People who have heart difficulties cannot respond quickly to the added stress to the system, which can cause problems for a sensitive or damaged heart.

Extreme drops in temperature can also induce asthma attacks in some people.

Excessive sunlight can also cause physical problems. In addition to the dangers of sunburn and possible skin cancer, in some people ultraviolet rays produce an inflammation in the inner eyelid (a form of conjunctivitis). This is an allergic reaction to the sun's radiation. In some cases, sneezing occurs.

At what temperature do most people feel and work their best?

There have been some studies conducted that suggest that a temperature of 75 degrees and a humidity of 55 to 60 percent, is perfect for the indoors.

Outdoors, this ideal figure can vary. If you're running a marathon, you would like the weather to be cloudy and in the 40-degree range. If you like to go swimming, obviously warmer water temperatures and sunshine is in order. Everyone has a favorite type of weather. A skier prays for snow, a gardener for a beautiful sunny day—with an occasional shower. Different strokes for different folks.

Can the weather give you a headache?

If you are extremely allergic to pollen, you live near a park or open area, and it is a windy day, it's conceivable that the spores or pollen in the air can trigger a reaction which may, in fact, produce teary eyes, a headache, or a bronchial condition.

Some studies indicate that violent swings of temperature and barometric pressure may affect people with a tendency toward migraines. However, much more research needs to be done in this area.

Did the weather have anything to do with the day you were born?

My mother had a lot to do with the day I was born. Seriously, though, there are indications that severe changes in barometric pressure may induce a quicker labor for a woman who is just about to give birth. If the barometric pressure drops rapidly, it seems to affect a woman in her ninth month.

Why do some people with arthritis or scars know what the weather will be before the weather forecaster does?

This may be an "old wive's tale" or folklore, but it does seem to contain a little bit of truth. When barometric pressure falls, water rises. Ninety percent of our blood is made up of water, so it would make sense that in people who have conditions such as scarring, blood flow becomes more difficult through that area. When a storm approaches and the barometer falls, the water in the blood rises and pushes against the scar tissue, and a person might experience some pain. Twelve to 24 hours before a storm, people who have scar tissue or arthritis often complain of an ache or an ankle hurting.

How does falling air pressure affect the human body?

If one goes too rapidly from high pressure to low pressure, especially in terms of altitude, one can feel pain in the arms and legs, or in muscles and joints, or experience what's commonly called the "bends." Sometimes there's fatigue, nausea, or vomiting.

You may have heard the term "mountain sickness." Mountain sickness occurs when people arriving at a mile or so above sea level, must quickly get to the lower levels of oxygen in the air. This may cause a slight palpitation of the heart, loss of appetite, or nosebleeds. In some cases, this condition makes it very difficult for people to exercise unless they acclimate themselves to this less-oxygenated air. For example, an athlete could easily run 5 or 8 miles at sea level, but if she goes to Denver, for example, after 2 or 3 miles she might experience some pain or tightness of breath. This would simply be caused by the lack of oxygen at high altitudes.

Is it tougher to get going on a rainy, cold morning—or is it just your imagination?

For most people, getting going on a rainy, cold morning should not be a problem. Studies show that it's usually not sun or rain that affects a person, it's the barometric pressure.

Do people generally feel their best during long periods of sunny, bright weather?

Again, if it's sunny and bright, with temperatures in the 50s, 60s, and 70s, along with intervals of some change, the human body adjusts very well to it. Studies have shown that during the long, gloomy days of winter and in long rainy periods, the body produces more of a hormone called malatonin. With an increase of this chemical in the body, people do tend to become more depressed.

It has also been shown that during long hot

stretches, tempers flare up and there is an increase in violence.

What are the best things to do during a heat wave or a cold spell?

During a heat wave, the temperature of the air is sometimes higher than the temperature of the body. If the air is very warm, heat from the body cannot escape. The trigger mechanism to regulate heat within the body is upset. The area of the brain that acts as a trigger mechanism or thermostate—very much like the thermostat of your house—is called the hypothalmus. Body heat then begins to increase, the pulse quickens, and the body produces further heat. This can lead to heat exhaustion or heat stroke. It's therefore important during a heat wave to stay in a cool place, drink plenty of fluids, and not to exercise outdoors except during early-morning or late-night hours. This is especially important for the elderly because their bodies may be less likely to tolerate extreme heat.

During a cold wave, just the opposite occurs. The capillaries closest to the skin begin to close down. This reduces the blood flow to areas of the body closest to the skin (it's the blood flow which is necessary to maintain the body warmth of the organs). Sometimes you may find yourself shivering or experiencing a quick, involuntary movement of the body. These are some of the ways the body com-

71

pensates for the loss of heat; the body is trying to warm itself. Usually, your body temperature falls no more than 1 or 2 degree from the normal 98.6 degrees. However, if the temperature change is so intense and the body isn't able to compensate—due to exposure, lack of clothing, or too much alcohol—further difficulties can occur.

During a cold wave, you may have a problem with your extremities—the nose, toes, fingers, ears—which if exposed too long, the layers of skin may freeze and you'll get frostbite (the freezing of the tissue of the skin). So, it's important in extremely cold weather to stay indoors as much as possible. If you are going out, wear at least a couple of layers of clothing (to trap body heat close to the body), and protect the extremities by wearing a hat and gloves, because the head and hands are where most body heat escapes.

Why does one's appetite increase when the weather turns colder?

When the body has to adjust to colder weather, the heart has to pump faster to provide blood at a more rapid pace. The body then must metabolize more rapidly—which simply means that more fuel is released. Your body can be compared to a fireplace roaring on a cold winter night. Food is the fuel for the body, the same way wood is the fuel for the fire. As more and more heat is needed to compensate for the cold, more and more food is needed as a fuel for the body to metabolize and provide the energy needed to counteract the cold from outside.

Are more babies born when there's a full moon?

There's not a lot of research to support this theory, but many people do believe it. During a full moon, tides can rise as much as 6 feet, sands can lift from the ocean as much as 3 inches, now, since 90 percent of our body is water, and a woman carries so much water in her pre-delivery stage, it's not impossible to believe that the pull of a full moon can somehow help induce labor.

Can your home be as dry as a desert?

Absolutely. A house or apartment can have a relative humidity of less than 20 percent—the same as it would be in the desert. Sometimes the skin will become scaly. It might even become difficult to breathe because it's so dry. You can reinforce relative humidity by putting a humidifier to work and getting a lot of evaporated water into the atmosphere of your home. A trick you can use if you don't have a humidifier is to put pots and pans of water on your radiator. Even worse is to have very dry air, loaded with pollutants, in your home. It irritates the mucous membranes in the nose, throat, and perhaps deeper into the bronchial tubes. It can also lead to mold spores in the apartment, which

73

would add a further drying effect in the throat and nose cavities as well as irritants that could cause some type of sinus or allergic reaction.

Can weather cause people to contemplate suicide?

There have been many studies done which have found that sunshine, gray skies, or rain conditions don't affect suicide rates. What did affect them (up to 30 percent, according to one study) was when the barometric pressure changed rapidly, between .25 and .50 inches in a 24-hour period. The study also showed that there were more suicide attempts when the barometric pressure was falling than when it was rising.

Can the weather affect how you spend money?

This is a very iffy question. There have been some studies that indicate that if you go to a restaurant on a sunny, mild day with low humidity, and you like your waiter, you'll probably tip him more. On days of cool, damp weather, people tend to tip less. What does all this mean? We're not quite sure. It might simply mean that stressful weather produces stressed reactions in people. I believe that these types of surveys really just reflect mood and the extra burden that bad weather can place on people, rather than an actual biological change which makes them behave differently.

74

6

Tomorrow's Weather: Ecologically Speaking

What's the difference between fog and smog?
Fog is, very simply, a cloud that settles to the earth.
It's condensed water droplets at very low levels. It
can settle on any part of a land mass, though it's
most common along the coastal areas—the east
coast and west coast. It's very natural, meaning it
occurs within the cycle of nature. Smog, on the
other hand, is quite different. Smog is man-made;
it's a combination of industrial smoke and fog.
When you have fog *and* smog, that can often be a
very dangerous combination. It can produce very
poor air quality. Eye, ear, throat, and lung irritation
can result. Smog is, of course, most common in cit-

Fog is really just a cloud that has settled on the ground. Here is a cloud that is trapped by a mountain and cannot move on. (Neg. No. 6356 fr. 10. Photo: R.E. Logan. Courtesy Department of Library Services, American Museum of Natural History)

ies, such as Pittsburgh, or places like Denver and New York City.

What is acid rain?

This unfortunate phenomenon is a product of the Industrial Age and the twentieth century. Acid rain is one of the most important environmental problems that we are facing right now and are going to face as we near the turn of the century. Tremendous amounts of pollutants, 50 million metric tons of sul-

76

phur and hydrogen oxide, are being poured into the atmosphere every year. Carbon dioxide, carbon monoxide, and burning gasoline, coal and other fossil fuels, through a series of very complicated chemical reactions, return to earth in the form of acidic rain and acidic snow. The impact is devastating. To cite just one example, many lakes in the Adirondacks can no longer support fish life. Acid rain is affecting lakes in not only the eastern portion of the United States but all in North America and northern Europe. Aquatic life is not the only life being affected. During the fall foliage season, you can see the effects of acid rain as well. Many leaves have holes burnt in them from falling rain.

Ever seen the stains on statues? Whether you're in the parks of Allentown, Pennsylvania, Kansas City, Missouri, or Boise, Idaho, or you're at the great historical monuments of Greece or Rome, you're looking at acid rain slowly eating away at the stonework and sculpture. Just like living creatures, man-made treasures are being destroyed by acid rain. And when they are gone, they are gone forever.

What is the ozone layer?

The ozone layer is a thin layer of an unstable form of oxygen. The heaviest concentration of this oxygen is very high up in the highest layer of the atmosphere, which is known as the stratosphere. It's thought that this unstabilzed layer of oxygen absorbs most of the ultraviolet solar radiation that the earth receives. (It's ultraviolet solar radiation that causes most cases of skin cancer.) This layer is thinning, and holes have been detected at the poles. Many of the same pollutants which cause acid rain are destroying the ozone layer. Another ozone-destroying villain is *CFC*, or chlorofluorocarbons, which are used in air conditioners and in some kinds of manufacturing.

What is global warming and the greenhouse effect?
The greenhouse effect or global warming is a phenomenon that probably began during the Industrial Revolution, but has only been documented as of 1958. The real problem is the tremendous increase in carbon dioxide (CO_2) in our atmosphere at all layers. CO_2 used to dissipate into the atmosphere or it was used up by the earth's greenery. Now fifty percent of the CO_2 released from fossil fuels has accumulated in the atmosphere and is very, very difficult to displace. By absorbing the long-wave radiations from the earth and not letting them go, CO_2 heats up the lower layer of the atmosphere. And as it heats up the lower layer of the atmosphere, that gas, becomes trapped and heating of the planet begins. It's possible that in the next 75 years we will double the amount of CO_2 output into the lower layers of the atmosphere, which could cause sub-

stantial warming. Even if the industrial nations curb emission levels from the burning of fossil fuels, which would control the distribution of CO_2, a lot of damage is already done. Countries which are still developing, such as Brazil and Peru, need to have incentives to curb emissions and to stop burning their forests.

If the atmosphere heats up uniformly around the planet, over a period of time changes could occur that would affect the weather. It's possible that the earth's temperature could raise a degree if emissions are not controlled. This could be disastrous. One consequence might be a slight melting of the polar cap. This could be very significant; if the caps melt, flooding could occur along the coastal sections.

Also, if the planet heats up, it's possible that the ocean would become warmer; hurricanes could become more frequent, and certainly more dangerous. Hurricanes now usually thrive in water temperatures of about 80 degrees, but if the water rose to 85 or 86 degrees, it would not be uncommon to have sustained hurricane winds of 200 miles per hour, and wave heights of 15 to 20 feet. Hurricanes could reach the catastrophic level—five—more frequently, causing massive damage to both life and property.

Scientists also theorize that climates would be altered. For example, if the average temperature went up 2 or 3 degrees, it would be possible that, over a period of time, New York City would have weather like that in Memphis, Tennessee, and that the northern plains of Canada would have weather like that in the Midwestern United States. Still, there would be areas, according to scientists, that would remain cold and wet; there would be pockets that would remain dry and drought-like.

It's difficult to predict what will really happen because all of this is based on theory. But scientists think the general trend is something like this: if global warming does occur, and if temperatures go up 1 to 3 degrees over the next 50 to 75 years, we'd see milder winters and our climate would become one of two seasons, cooler and warmer, rather than the four seasons that we're accustomed to. But you have to remember that these are computer-based models; it doesn't mean it's definitely going to happen.

It should be noted that over the last hundred years in the United States, the average temperature has not gone up all that much. It has gone up less than 1 degree, which statistically is not that impressive. But the amount of airborn pollutants caused by the uncontrolled burning of fossil fuels around the planet suggests that somewhere along the line, these predictions will actually come true.

The biggest problem is convincing people that global warming is something to be concerned about. Global warming, or the greenhouse effect, is a long-term problem, though there's an argument made by some scientists that it really isn't a problem at all, that it's been sensationalized in the media. But you should be aware that the burning of CO_2 around the world continues, that Mother Nature is beginning to say to us for the first time that perhaps the cycles have been interrupted.

What is the heat-island effect?

The heat-island effect is usually found in cities that have a lot of towering concrete buildings. During a hot summer day, heat begins to build up and is trapped by the "concrete canyons" (which actually sweat). Surface temperatures on the ground can be around 90 degrees. However, the buildings tend to heat up the surface temperature even more—4 or 5 degrees higher. You might have felt this at a ball game inside a stadium. The temperature outside may be 95 or 100 degrees. But when you take the concrete which traps heat, you take sunlight and the lack of wind (due to the trapping of air in the "concrete canyon"), the temperature can soar some 5 or 10 degrees above the surface temperature. Those ball players have to sweat for their money!

Heat-island effects can also cause terrific thunderstorms, which build up and cross many large cities. As a matter of fact, if such a thunderstorm stalls over an area—Chicago, New York City, or Pittsburgh, say—a torrential amount of rain can fall in a short period of time. And since big cities are expanding, the effect is occurring in more and more places.

What is the difference between a drought watch and a drought warning?

Both a drought watch and a drought warning are indications of problems with reservoir levels (supply of water). When the difference between normal average levels and actual levels drops beyond 15 percent, a drought *watch* is declared. When the difference becomes more than 20 percent, a drought *warning* is issued. Then conservation is the word—and use of water should be watched. Generally, guidelines are offered. Restrictions and rationing are not enforced. Ways in which people can conserve water include not serving water in restaurants, not washing cars as much, not watering lawns except for short periods of time, limiting municipal-pool use to certain times, and taking shorter showers.

Why are droughts common in the United States?

Droughts are not only common in the United States, but also in Canada, Europe, South America, Africa—in short, all over the world. Thereare two ways to look at drought. First, they are cyclical in nature: they come and they go. There are years of plentiful rainfall and years of little rainfall. There are years of drought and years without drought. It seems that the areas of the northeastern part of the United States and also the west coast are experiencing more drought conditions. The big question is whether droughts are part of a trend toward global warming and the greenhouse effect or whether they are just the shifting of the normal cycles of wet and dry.

So the question really becomes . . .

Are more droughts part of our weather future?

There are some indications that the cycles of drought and rain are being affected by global warming. This warming, though very slow, will begin to shape warming trends if the burning of fossil fuels continues uncontrolled. Scientists are theorizing that, in the next 10 years, there will be a trend much more toward dry, warm weather.

If these indications are correct and the burning of fossil fuels continues uncontrolled—and thus, global warming continues—then the reality may be that, 15 to 20 years from now, dryness will become more pronounced. Droughts will be longer, more sustained over larger sections of the land than in years past. If global warming is *not* as pronounced—if the planet does not warm up more than one-half of one degree, then these situations will not come to pass, and we'll simply witness a cyclical drought cycle. However, there *are* indications that we are tampering with the environment on a scale not previously known to man.

Is there concrete proof of global warming?

With all the research being conducted on this issue, the answer to this question is still no. Most of the burning of fossil fuels has occurred primarily in this century, so the the long-term data is suspect. The chances are great that global warming has been a gradual process since the Industrial Revolution. But because this heating is a cumulative action, between 1990 and the year 2000 we may see warming unparalleled in human history. If there is warming in the next 10 years by a 1 degree, the earth will have warmed more since 1958 than it has in the last thousand years.

We may be in the embryonic or the infancy stage of global warming and statistically not able to show it yet, but once we *can* prove it, it may be too late. So it's better that we start limiting emissions now and take a good look at what we're doing to our environment.

What is an ozone alert and an ozone warning?

An ozone *alert* is the *possibility* of dangerous levels of ozone in the atmosphere. It's suggested that people with respiratory diseases take caution with their activities.

An ozone *warning* means that conditions are unhealthy: the levels of ozone are too high, and people with respiratory or lung diseases or asthma should not be exercising, and should stay indoors where the air quality is better.

Why, when it rains, are there still reports that the reservoirs are so far below their normal capacity?
One inch of rain can lift the capacity of a reservoir anywhere from 3 to 5 percent if it's a direct rainfall—but that's *only* if it rains directly into the reservoir. And with snow melting off into the reservoirs, that ratio would remain the same. The problem with the reservoir systems is many of them are very old, and with an old system there are problems that develop. One of the problems is leakage. For example, in the New York reservoir system there are 560 billion gallons of water stored. Unfortunately, 7.5 percent of the water which is piped every day from the mountains hundreds of miles to New York City doesn't make it through because of the hundred or so miles of leaky pipes. That's a tremendous amount of water that's lost every year. Actually, you could argue that normal rainfall into the reservoirs is not really enough, that there needs to be a 7.5 percent-above-normal rainfall just to compensate for the leakage. This is another difficulty; you can't count on having normal rainfall or above-normal rainfall every year.

7

Weather-Forecasting Groundhogs and Other Folklore

Can a leech forecast the weather?

Believe it or not, long ago, it was thought that the leech—a type of worm—had great forecasting skills. In the nineteenth century, a British doctor put a leech down at the bottom of a bottle. He gave it air and attached it to a bell. What happened was that during fair weather (blue skies, calm barometric pressure), the leech hardly moved. However, when there were periods of barometric change and storms, the leech rose to the top of the bottle (thus causing the bell to which it was attached to ring). The faster it would move to the top, the harder it rang the bell, and the more severe a storm would be. So, if you don't believe your local weatherman, get a worm, stick him in a bottle, and hook him up to a bell—and compare the forecasting records of the weather-person and the weather-worm!

Red sky at night, sailor's delight.
Red sky at morning, sailors take warning.
True or false?

There is a Biblical expression that says, "When it is evening ye say it will be fair weather for the sky is red. And in the morning it will be foul weather today for the sky is red and lowering." Let's see if we can verify these expressions. Look at the sky in the morning. If that's too early for you, test the sky at night. If you see red sunsets, it generally means the air is dry and there's no cloud cover; you'll probably have a fair day coming up. If the sunrise is red, it generally means that there's dry air, but clouds could move in from the west.

A fiery sunset, lasting well after the sun has moved below the horizon, usually means there's a low-pressure system and that the rays are reflecting off distant clouds. So, while this old saying has some truth to it, there is, of course, an exception to every rule.

Can you tell the temperature by a cricket's chirp?

This is a very tricky question. It actually may work if you are good cricket-chirp counter. Here's how you would do it: by adding the number of times a cricket chirps in 14 seconds to the number 40, you will come very close to the surface temperature outside. The question then is: which is better, a good thermometer or a good cricket?

Does March really come in like a lion and go out like a lamb?

Well, no. March usually is blustery in the beginning and can be very spring-like at the end. It really is an expression to show that March is a month that shows very sharp contrasts.

What is meant by the dog days?

The "dog days" expression is heard very frequently during the hot days of summer.

The star known as *Sirius* (*sirius* is Greek for dog; the constellation looks like a dog) which is a star in the constellation Canis Major usually appears in the summertime. (So the "dog days" could come from the fact that the constellation shows itself during the hottest days of the summer.)

Onion skin very thin, mild winter coming in.
Onion skin thick and tough, coming winter cold and rough.
True or false?

During an onion's growth, the amount of moisture available to it will determine the thinness or thickness of its skin. So this saying has more to do with the amount of moisture that's available while an onion is growing than with its ability to forecast long-range weather patterns.

If ants their walls do frequent build.
Rain will from the clouds be spilled.
True or False
We cannot use the strength of an ant wall for long-range weather forecasting. Ants building their walls, muskrats building their houses, beavers building their dams—all build based on the amount of water or moisture available. If there's not enough moisture to build a firm wall, ants will build loose walls again and again, until the structure is able to hold steady.

Why is a rooster used on a weathervane?
The rooster was first used on a weathervane in the ninth century because of an order by the Pope. In the book of Luke in the New Testament Jesus says to Peter "Before the rooster crows, you will deny me three times." The church chose the rooster to be the symbol for St. Peter.

Halo around the moon.
Rain or snow to come soon.
Usually what happens is that the sun reflects off the moon, and the moon reflects off ice crystals in high, thin cirrous clouds over earth. Those crystals will produce the halo-like effect appearing around the

moon. It just so happens that those cirrous clouds are the vanguard of a storm. The storm may or may not hit, but the halo effect usually is a sign of thickening clouds and possibly some precipitation. There's a bit of truth, then, to this expression.

Buy a house before it rains or snows and I'll save you a fortune.

This one is especially interesting. It can perhaps save your family money! Before a storm arrives (24 to 48 hours ahead of time), air tends to rise. If there's a garbage dump miles away, you may not pick up the odor on a bright, sunny day. But before a storm arrives, odors from the land are lifted and you can smell odors from miles away. A real-estate agent may not want to tell you of the proximity of a garbage dump, but being a little weather-wise can sometimes help money-wise.

Squirrels gathering few nuts means an easy winter, squirrels gathering lots of nuts means a tough winter.
True or false?

This bit of folklore goes back centuries. Squirrels gathering nuts has absolutely nothing to do with how harsh or mild the winter will be. It has a lot to do with how many nuts are available to be harvested.

Groundhog Day: who do you believe, the rodent or me?

According to tradition, if, on February 2nd, the groundhog sees his shadow, there will be 6 more weeks of winter. If the groundhog does not see his shadow, spring will arrive early. The tradition of predicting spring on February 2nd really comes from the celebration of itself—Candlemas. There is an English proverb that says: "If Candlemas be fair and bright, winter will have another flight, but if it be dark or clouds of rain, winter is gone and will not come again." The shadow cast by the groundhog has nothing to do with the original folklore that Candlemas itself was a predictor of the weather. Sorry, hog lovers.

What does "It's raining cats and dogs" mean?

This saying comes from the German expression "It's raining cats and ducks." Cats, it's said, like to go indoors when it rains and stay by the fireplace and cuddle and huddle, and ducks like to enjoy the rain because they can use their webbed feet to play in the puddles. So a hard rain chases cats inside and ducks outside. "It's raining cats and dogs" probably is a weather belief and a weather extraction from the German saying.

When ants travel in a straight line, it will rain soon. True or false?

Absolutely false. [Ants don't travel in a straight line.] If you want to put this to the test, just put some ants in your apartment and see what they do before it rains. Chances are they will run all over the place. I wouldn't recommend this, nor would I recommend the other legend, "Step on an ant and it will rain soon." The answer is: false.

Thunder curdles cream, lightning sours milk. True or false?

Before refrigeration was invented, people kept milk in cellars—and lightning couldn't reach in there, of course. If cream curdled, and milk began to taste bad, it wasn't because of lightning; it was probably caused by the higher temperatures that often occur before thunderstorms.

A cow with its tail to the west makes the weather best, a cow with its tail to the east makes the weather least. True or false?

When you see a herd of cows, they are often all facing the same direction. Cows tend not to like their face in the wind, so if you see the tails pointing to the west, backsides to the winds, faces out of the wind it means that there will be fair weather. Westerly winds are the prevailing winds—the dry winds which bring good weather. So there may be just a bit of truth to this expression: when a cow's tail is facing to the east, that usually means more fair weather is coming, indicative of counter-circulation of a storm, or winds off the water or ocean.

What do torrential storms and bananas have in common?

The vegetation on many tropical islands desperately needs the rain that's brought by hurricanes. But with the storms come winds that can ruin crops. Bananas are one of the only crops that can survive such storms. And that's no monkey business!

How can a tree tell us about the weather hundreds of years ago?

Each ring in a cut tree shows a year's growth. If the rings are close together, it was a cold year. But if the rings are far apart, then—you guessed it—the year was warm.

Is it true that when cows lie down, it will rain?
Some people believe that cows can sense moisture in the air and know when storms are coming, so they lie down in order to have a dry spot and be sheltered from the storm. Studies show that the cows are wrong more often than not.

If the oak flowers before the ash, we will have a splash.
If the ash flowers before the oak, we will have a soak.
True or false?
This is an old English saying. The first line means that there will be light rain and mild weather for the next month or so. The second line predicts a month or so of wet, gloomy weather. Again, there's not much evidence to support this one.

Can you tell the weather from a pinecone?
Here's one weather-forecasting trick that does work. If rain is coming, a pinecone will close up tight. If the weather is dry, a pinecone's scales will dry out, shrivel, and pop out, giving it the look we're most familiar with.

8

How Do You Become a Weather Forecaster?

How can you discover early if you're a weather nut?

Well, in my case it was really quite simple. When I was 8 years old, if it was cold outside, with thickening clouds, I would run home, turn on the radio or television, read the newspapers, and open the window to wait for the first snowflake.

If you're interested in finding out about—by touching, smelling, listening—anything involving weather, you are a weather buff.

What should you study in school to become a weather forecaster?

There are many different routes to becoming a fore-

95

caster, but the basics include a good general background in math and the sciences, especially physics. Hopefully, you'd get a degree in meteorology and/or communications. It's important to read, read, read all kinds of books—and to be able to express yourself clearly. It's also important to be able to work cooperatively and enthusiastically with other people.

What does the National Weather Service do?

The National Weather Service's main purpose is to provide the U.S. with forecast data—information which is distributed throughout the country.

The National Weather Service really proves its worth when forecasting severe storms. The Service has the responsibility of issuing watches and warnings for storms, tornadoes, hurricanes and snow and other winter weather, as well as reports on road-travel conditions and marine interests (that is, gale and surf conditions). The National Weather Service has a vital role in providing the public with a huge amount of information so that we can make decisions about our day, our well-being, our safety, and, in instances of extreme weather, our lives!

As much as possible, the Service must be quick, fast, and accurate in responding to weather conditions throughout the United States.

Where do television and radio stations look for their weather forecasters?

Weather forecasters usually come out of communications schools. They search out jobs themselves, preparing demo tapes and starting out in smaller TV markets.

The best route, I think, is to intern at a major station. Watch how the pros do it, and then get your feet wet at a smaller station. You might enter the marketplace by first doing an early-morning show, then getting to the main show. (Do stories, learn how to produce, and shoot pieces.)

Being a weather forecaster involves more than simply displaying one's weather expertise. There are certain qualities that you must have, whether it's charm, personality, credibility, a soft smile, a warmth, a reaching out. I call them all natural talents—you either have them or you don't. Anything forced on television just doesn't work. So a very "natural" person is the one who will get the best jobs. Even though there are many weather-people, there are very few who have that natural, easy ability to communicate comfortably, with a friendly smile and in a believable way.

What are some of the fields a meteorologist can get involved in?

You could teach meteorology at the high-school or

university level—especially at a modernized high school. You could work for environmental services. You could work for the state or federal government. You could, of course, work in radio or television. Or you could work for the national oceanic (NOAA) or space administrations (NASA).

What are the job prospects for meteorologists?
Job prospects in the government seem to be diminishing because of cutbacks at the National Weather Service. The private sector looks better. Many small weather-oriented business services are popping up which cater to television and radio stations, newspapers, all kinds of business (agriculture, tourism, fashion), airlines, and marine interests.

What do you do if you are gainfully employed but want to study weather and become a meterologist?
I get many letters from people in other branches of business who would love to get into meteorology and want to know how to get started. The answer is, it's a tough move. Going from accounting, say, to meteorology, you'd have to do things someone just starting out would do: intern with a professional, read a lot on your own, perhaps go for a degree part-time, or at least satisfy requirements so that you know enough to try your hand at a radio or television broadcast. Above all, you should keep your foot in the water and keep moving. You never

know what can happen. Many meteorologists and other weather-people don't have formal degrees but are quite good at what they do.

Are there quality weather stations that you can set up in your house?

Yes. Check the catalogs. The Talyor Corporation makes a lot of quality instruments that both look good and function well.

There are indoor weather stations and outdoor weather stations, and even computer programs that you can buy from private weather forecasting companies to set up on your personal computer. Those programs can provide live information to you, which is very exciting because you then have some of the most important tools a weatherman has—satellite information, some of the things that make weather so interesting and so vital.

Brief Bibliography

Branley, Franklin. *It's Raining Cats and Dogs.* Boston: Houghton Mifflin,1987.

Cosgrove, Brian. *Weather: An Eyewitness Book.* New York: Alfred A. Knopf, 1991.

Lockhart, Gary. *The Weather Companion.* New York: John Wiley & Sons, 1988.

Thompson, Philip D., Robert O'Brien, and the editors of *Life. Weather.* New York: Time, Inc., 1965.

IRV "MR. G" GIKOFSKY was a teacher for eleven years before becoming a weatherman. He is currently the national weatherman for CBS radio network and, until recently, was the weatherman for Channel 2 in New York City where he lives.

STRANGER THAN FICTION

by MELVIN BERGER

ASTOUND YOUR FRIENDS
WITH INCREDIBLE, LITTLE-KNOWN FACTS ABOUT...

KILLER BUGS   76036-3/$3.50 US/$4.25 Can

More people are killed by insects than by all other animals combined—including sharks and snakes.

DINOSAURS   76052-5/$2.95 US/$3.50 Can

Dinosaurs are the largest, most magnificent and most terrifying creatures that ever roamed the Earth.

MONSTERS   76053-3/$2.95 US/$3.50 Can

Do creatures like Big Foot and the Abominable Snowman really exist?

SEA MONSTERS   76054-1/$2.95 US/$3.50 Can

Unlike the shark in *Jaws*, this book is about the real living sea monsters that swim the waters of the world.